NATIONAL GEOGRAPHIC

Ladders

BIG BEND National Park

GENRE Social Studies Article

Read to find out how and why Big Bend National Park was created and preserved.

Strap on your backpack, find your sunglasses, and start exploring Big Bend National Park. This park offers many different types of landforms, including mountains, canyons, and deserts. What forces shaped this rugged land? Over hundreds of millions of years, ancient oceans, earthquakes, and volcanoes formed and exposed the park's rocks. Wind and water then carved the peaks, canyons, and deserts we see here today.

WELCOME TO BIG BEND

by Cynthia Clampitt

Big Bend lies in the Chihuahuan (chih-WAH-wahn) Desert in southwestern Texas, on the Mexican border. In fact, Big Bend is named for the sharp turn, or "big bend," a river called the Rio Grande takes as it flows between the United States and Mexico. Most people think of deserts as hot, harsh places where few plants and animals can live, but that's not true of Big Bend. Yes, Big Bend can be hot and harsh, but the park protects several unique desert **habitats**. It is one of the best places in the United States to see desert wildlife—in fact, this land has been packed full of living things since long before dinosaurs roamed here.

Big Bend's mountains are a challenge for climbers.

A TREASURED NATIONAL PARK

The first people to live in this region were Native Americans who arrived about 10,000 years ago. Thousands of years later, the Spanish came looking for gold and other riches, and miners came searching for minerals to pull from the earth, too. They didn't find gold, but they found rocks that contained the metals silver and mercury.

The real treasures of Big Bend, though, are its beautiful landforms and plentiful wildlife. To preserve these treasures, Texas set aside this area as Texas Canyons State Park in the 1930s. In 1944, the park was turned into a national park and renamed Big Bend. But Big Bend is more than a national park. To help protect and preserve this fragile **ecosystem** even more, Big Bend is also a biosphere reserve.

Big Bend bluebonnets bloom on the banks of the Rio Grande.

A **biosphere** is the combination of earth, air, and water that supports life, and a reserve is a place set aside for study. The Chihuahuan Desert, where Big Bend is found, is a biosphere rich in wildlife and special landforms, so the park was named a biosphere reserve in 1976. Scientists study this biosphere carefully to learn how to protect an even larger biosphere—Earth.

Big Bend's biosphere includes more than just living creatures—in fact, some of the park's most famous creatures have been dead for millions of years! Thousands of **fossils** have been found here, including fossils of gigantic crocodiles and pterosaurs (TAIR-uh-sawrz), or flying reptiles. Researchers have found fossils of more than 90 types of dinosaurs at Big Bend, such as the bones of an enormous creature called Alamosaurus (AL-uh-moh-SAWR-uhs).

THE MAKING OF BIG BEND

1903 Cipriáno Hernandez builds a farm and store near the Rio Grande. The area grows into the town of Castolon.

1944 Big Bend becomes a national park to preserve and protect its unique landforms and wildlife.

1964 NASA's Apollo astronauts visit the park. They study volcanic rocks as they prepare to go to the moon.

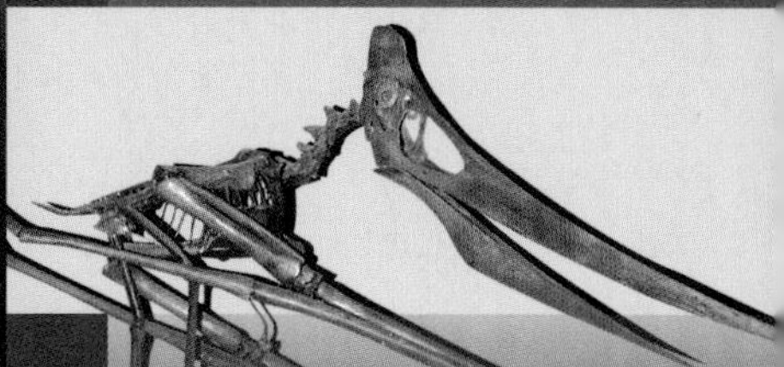

1971 A college student finds fossil pterosaurs, or flying reptiles. They are some of the largest ever found.

2011 About 360,000 people visit Big Bend to hike, bike, canoe on the Rio Grande, and enjoy the wildlife.

THE WILDLIFE OF BIG BEND

Today, Big Bend is home to all types of living things. Flitting hummingbirds, swift roadrunners, soaring eagles, and many more kinds of birds live here—more than 450 different species. But the bugs beat out the birds. Over 3,600 kinds of insects can be found in Big Bend. It's a bug paradise! All of these animals have adapted, or changed, to live in the Chihuahuan Desert.

JAVELINA

The javelina (hah-vuh-LEE-nuh) isn't a pig, but it sure looks like one. In Big Bend, javelinas usually travel in herds of about 14 animals. Other mammals in the park include mountain lions, black bears, bobcats, and coyotes.

How can this park support all this life? Big Bend has mountains as well as flat desert lands. The temperature is cooler in the mountains, and there is a greater chance rain will fall there. In fact, Big Bend's mountains get twice as much rain as its desert lands, and this creates many habitats where certain plants and animals can live and grow. Let's take a look at a few of the creatures that live in Big Bend.

TARANTULA

A tarantula is one big spider. Some Big Bend tarantulas grow to be larger than the palm of your hand. But don't worry—tarantulas rarely harm humans.

COLLARED LIZARD

Big Bend's warm desert habitat is perfect for reptiles. This collared lizard can grow to be more than a foot long, and it can run on its hind legs.

LUBBER GRASSHOPPER

The lubber is a giant grasshopper that can grow to be three inches long. This insect's bright colors warn predators not to eat it. It's poisonous.

Check In What makes Big Bend different from other national parks?

Read to find out about the birds of Big Bend National Park.

Welcome to Big Bend, fellow bird-watchers! You've made it this far, so you know that this park is in a **remote** area of western Texas. It's so far from civilization that it's one of the least-visited national parks in the country. That's great for bird-watching because with so few crowds, our feathery friends feel safe to fly around the park. In fact, Big Bend boasts more birds than any other national park.

Let's visit three different habitats, places where animals and plants live naturally, that are preserved in this national park. First, we'll check out the birds of the Chisos Basin, high up in the mountains near the center of the park. Then we'll hike across the desert to the Chimneys, where birds nest near tall rock towers, and we'll end up on the banks of the Rio Grande, where the river has carved deep Santa Elena Canyon.

Birds aren't the only creatures that flock to Big Bend—scientists flock here, too, to learn about Earth's history. Hundreds of millions of years of volcanoes, erosion, and earthquakes have pushed up and worn down the land in the park, resulting in the canyons, deserts, and rock formations that provide homes for the park's 450 bird species. Grab your binoculars—let's check out the birds of Big Bend!

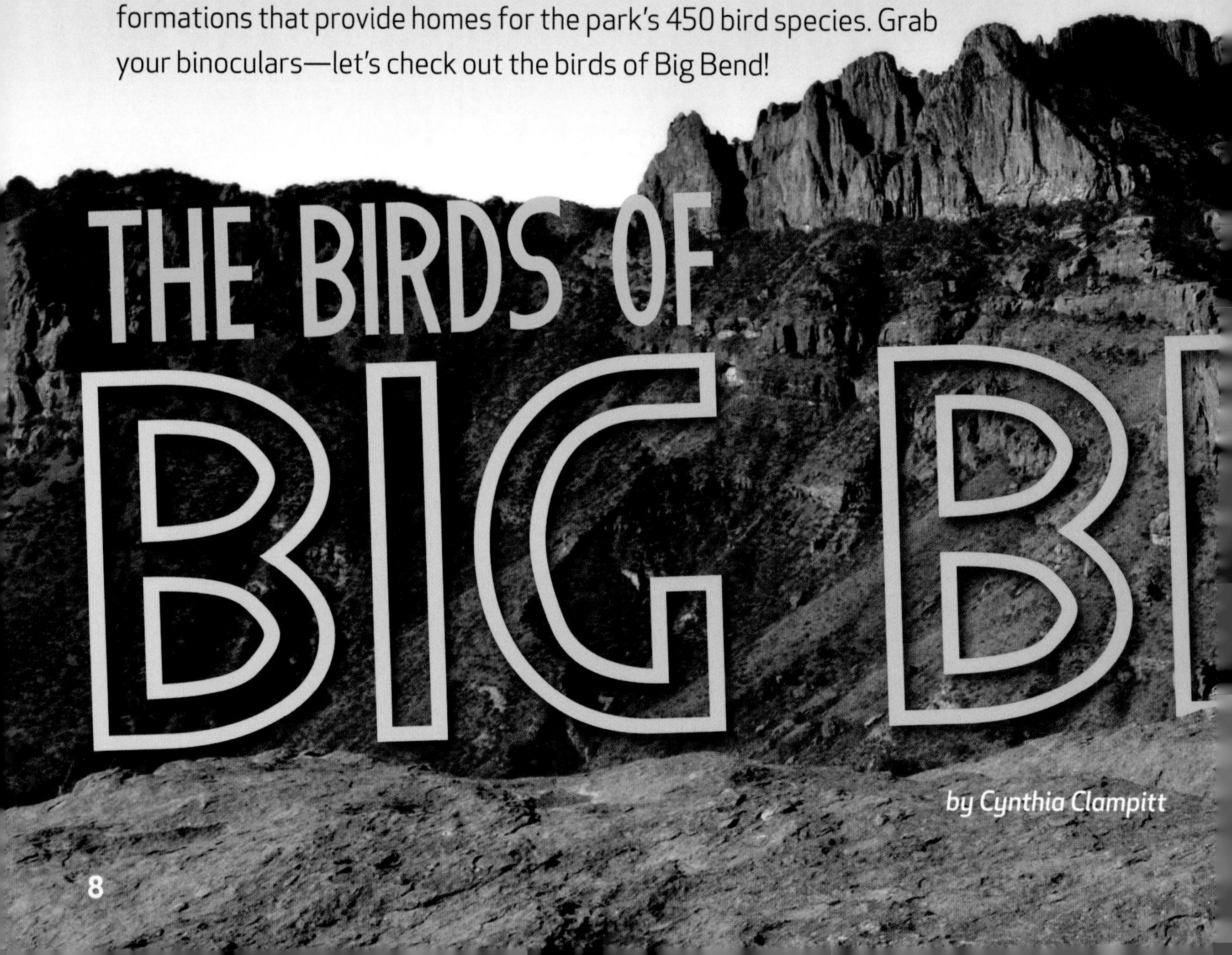

THE BIRDS OF BIG B

by Cynthia Clampitt

BIG BEND'S BIRDING TOUR

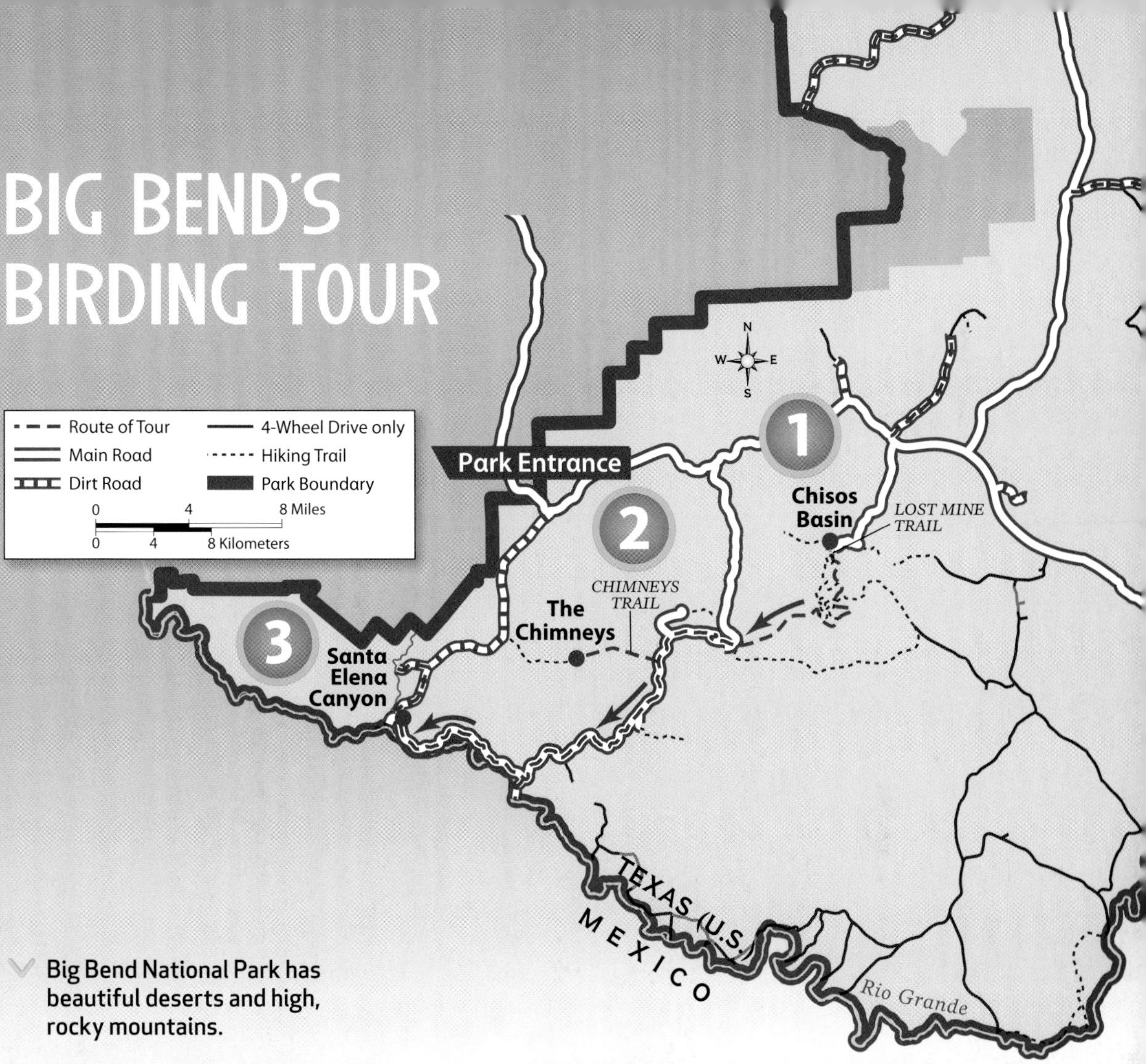

Big Bend National Park has beautiful deserts and high, rocky mountains.

1 Chisos Basin: Mountain Birds

Our first stop is Chisos Basin. A **basin** is a low area drained by a river, but Chisos Basin is high in the Chisos Mountains. It's just lower than the peaks around it.

Let's hike the short Window View Loop Trail. The Window is a high canyon at the edge of the basin. Water flows from the mountains through the Window and into the Chihuahuan Desert. It then flows to the Rio Grande, many miles across the desert. Keep an eye out for birds along the trail—there are plenty up here, but most of them try to blend in with the scenery. It may be hard to catch a glimpse of some of them.

The mountains are cooler than the desert below, so they attract many different types of birds that nest in the pine, oak, and juniper trees that grow up here. This is a good habitat for all kinds of birds, from tiny hummingbirds to magnificent **raptors**—birds of prey, such as eagles.

If you hear noisy chattering, you'll know that an angry blue Mexican jay is scolding us for coming too close to its perch on a juniper tree. If you're very patient and quiet, you might see a colorful Lucifer hummingbird sipping from the flower of an agave, a desert plant. And don't forget to look up! You may spy a golden eagle soaring on the desert wind.

It's not easy to spot these birds, so check out the "gallery" of bird photos on the right to see which ones you may have missed.

Lucifer hummingbirds build their nests on steep, rocky slopes, such as those in the Chisos Mountains.

As you can guess, most Mexican jays live in Mexico. However, many of these noisy blue birds also live in Big Bend National Park.

Golden eagles can dive through the sky at speeds of more than 150 miles per hour to catch their prey.

Roadrunners would rather scurry along the floor of the desert than fly above it. They can run more than 18 miles per hour.

The cactus wren nests in cactus plants and gets its water from the food it eats. It has adapted well to desert living.

The canyon towhee is a camouflage artist. Its brown and rusty feathers blend in with the desert floor to hide it from predators.

2 The Chimneys: Desert Birds

Next, we'll hike through the desert to the Chimneys, keeping a close watch for desert birds along the way. The Chimneys Trail leads straight to those tall rock towers in the distance. It's flat and hot, with no shade, so put on your sunblock, put on your hat, and let's go. Look! There's a roadrunner scurrying along the trail ahead of us. Roadrunners don't fly very well, so they usually hunt for snakes and lizards on the run.

As we hike, keep an eye out for the cactus wren. It builds its nest in cactus plants and gets all its water from the insects and spiders it eats. That's good because it won't find many other water sources out here. Listen—hear that cheerful chirping? It sounds like "chili, chili, chili." That's a canyon towhee. These brown birds are easier to hear than to see because they blend right into the desert landscape.

As we get nearer to the Chimneys, you might be wondering what they are doing out here in the middle of the desert. They formed when volcanoes erupted here long ago. Melted rock pushed up through cracks in Earth's crust and hardened to form veins of volcanic stone. Then the softer rock around these veins wore away, leaving the volcanic stone standing like towers or chimneys.

3 The Rio Grande: River Birds

Our last stop is the Rio Grande, the river that forms part of the border between the United States and Mexico. To get a good look at river birds, we'll paddle canoes into Santa Elena Canyon, which boasts the park's tallest cliffs. They soar as high as 1,500 feet straight up out of the river. That's as tall as five football fields stacked end to end!

As we pass through the trees on our way to the riverbank, look for holes in the bark. You might see the eyes of a tiny elf owl peering out at you. These little owls find holes in trees and cactus plants where they make their nests, and they fly at night to hunt insects and spiders near the river.

Now that we're on the water, keep an eye out for great blue herons wading in the shallows. The Rio Grande is a perfect habitat for these tall, stately birds. They keep very still as they search for fish, frogs, lizards, and small mammals to eat.

Those flashes of red high on the cliffs are vermilion flycatchers hunting insects as they fly. They nest in the trees along the riverbanks and atop the canyon walls.

The sun is setting over Big Bend, and our day of bird-watching is over. We saw lots of different birds, but there are hundreds more to see here. Come back anytime, and bring your binoculars!

Only male vermilion flycatchers are red. Females are brown. The males like to show off for the females by bringing them colorful butterflies to eat.

Ever heard a bird that sounds like a dog? Meet the elf owl, which sounds like a poodle yapping. While unusual, its calls are a common sound during nights in Big Bend.

The great blue heron is a wading bird, which is why it has such long legs. It is the largest wading bird in North America. It can grow to be four-and-a-half feet tall.

Check In How are the birds of Big Bend suited to live in their different habitats?

GENRE Legend **Read to find out** about the legend of the Lost Mine.

In Search of the LOST MINE

by Elizabeth Massie *Illustrated by Jan Lieffering*

THE LOST MINE TRAIL, A FAMOUS HIKING TRAIL THAT BEGINS NEAR THE CENTER OF THE PARK, CLIMBS HIGH INTO BIG BEND'S CHISOS MOUNTAINS. IT'S NOT A LONG TRAIL, ONLY ABOUT FIVE MILES UP AND BACK, BUT PARTS OF IT ARE STEEP AND QUITE RUGGED. THE TRAIL IS NAMED AFTER A LEGENDARY MINE BRIMMING WITH GOLD. IS THERE REALLY A LOST GOLD MINE IN THE CHISOS, OR IS IT JUST A LEGEND? IF THIS MINE ACTUALLY DOES EXIST, WHERE MIGHT IT BE, AND WHAT ARE THE ODDS THAT SOMEONE MIGHT FIND IT ONE DAY?

Early on a sunny autumn morning, the Jordan family laced up their hiking boots, filled their water bottles, and set off on their trek. Mom, Dad, eleven-year-old Paul, and seven-year-old Angela were vacationing in Big Bend National Park, and they had been looking forward to a long hike in the mountains. Paul had read about the "Lost Mine Trail" in the park brochure, and now the whole family was curious to explore it.

As they hiked upward into the Chisos Mountains, the family made a game out of identifying all the different lizards, birds, and plants. Paul pointed to a bristly tree and checked his handy park brochure again. "That's called an alligator juniper. Its bark is scaly, like alligator skin."

"And look, there's a deer!" said Angela. A large deer was grazing near the trail. Realizing it had been spotted, it glanced nervously at the family, swished its white tail, and bounded away.

Paul tugged at the brim of his baseball cap, pulling it down to shade his face from the hot sun, and wondered aloud, "If it's called the Lost Mine Trail, do you think there's really a mine hidden around here? Better look out, Ange, or you could fall right into it," he smirked.

Just then, a hiker headed down the trail toward Paul and his family, holding a long walking stick in one of her hands and a water bottle in the other. She smiled, raised her hand in greeting, and looked at Paul. "If I heard correctly, young man, you want to know if there's really a mine around here." Paul nodded, curious to hear what she knew about the lost mine but a little embarrassed that she had heard what he said to his sister.

"If you have a few minutes, I can tell you all about it," said the woman. She sat down on a large boulder, took a drink from her water bottle, and began her tale. "It's an old legend that goes back to the days when Spain ruled this land. Nearly 500 years ago, Spanish explorers sailed to North America hoping to find gold and other riches. The legend tells that there was gold to be found in these mountains, lots of it, and the Spaniards found it."

"The Spaniards claimed this land, built a secure fort and a prison near here, and called it ***Presidio de San Vicente***, which means 'Saint Vincent's Fort.' They had come to this region armed with powerful weapons, and they used these weapons to capture many of the Native Americans who lived in the land around Big Bend. The Spaniards kept their captives locked up in the prison within the fort."

"Every day, the Spaniards forced their prisoners to hike up into the mountains and work in the gold mine. Each time the prisoners left the fort for the gold mine, the guards blindfolded them and made sure those blindfolds were good and tight. The Spaniards did not want their prisoners to see where they were going. That way, they would never know how to get to the gold mine. And if they didn't know how to get there, they wouldn't be able to reveal its location."

"Mining was backbreaking labor to begin with, but the Spaniards would never let the prisoners take off their blindfolds," the woman continued.

"They had to work in blindfolds? How could they see what they were doing?" asked Angela.

The hiker nodded. "Most times they couldn't see, so they just did their best. But every now and then, some of the prisoners could peek beneath their blindfolds. They began to figure out the location of the mine and think about escaping. If they knew where the mine was, maybe they could come back one day and take some of the gold for themselves. But their cruel guards were very watchful. Escape was not likely."

"One evening, no prisoners or guards returned to the Presidio. The Spaniards waited impatiently, shading their eyes as they gazed at the mountains, pacing, and watching. Yet they saw no one."

"The Spaniards were furious. Had their prisoners and guards plotted together to steal the gold? They decided to hunt down the prisoners and guards and make them very sorry."

"Then what happened?" asked Paul anxiously.

The hiker continued. "The angry Spaniards shouldered their weapons and set off into the mountains, but they soon lost their way. They hiked through the long, hot day and into the cold desert night, but they never found the gold mine, the guards, or the prisoners. In fact, they were so lost that they never found their way back to Presidio de San Vicente."

"Weeks, months, and years passed, but the Spaniards didn't return from the Chisos Mountains, and neither did their prisoners nor the guards. No one knows what happened. We just know that every last one of them disappeared, and so did the gold mine and all its riches. To this day, no one has found it, even though many people have tried."

"What are we waiting for? Let's find that mine and get our hands on some of that gold!" said Angela, with visions of dollar signs dancing through her head. "Are there any clues about where it might be?"

"Well," said the woman as she rubbed her chin thoughtfully, "The locals say that if you stand on the site of the old Presidio's doorway on a certain day in spring and look up into the Chisos Mountains, the sun's first rays will strike the hidden doorway to the long-lost mine."

"Wow. Isn't that amazing, Paul?" Angela chirped, excitedly poking her brother in the chest, much to his irritation.

"Who knows if the mine ever existed at all," shrugged the hiker, getting up to leave, adding, "I'm sure it's nothing but a legend." She wished the Jordan family a pleasant hike and headed down the trail, treading carefully, walking stick in hand. Seconds later, Paul and Angela were leading their parents back up the trail. All of them watched carefully for clues to the hidden gold mine as they hiked, dreaming of the fortune that might be buried somewhere deep in the mountains.

"Check it out!" exclaimed Paul a few minutes later, pointing to a small gap in the rocks far off the trail. "That looks like a hidden pathway to me!"

"Sorry to disappoint you, son," said Dad, taking a closer look, "but I think that's just a spot where rocks may have worn down or fallen away."

Suddenly, Angela stopped and reached for a glittering rock by the side of the trail. "I found some gold! It must be from the Lost Mine!"

Mom smiled as she examined Angela's discovery. "Honey, that's just iron pyrite. Miners call it 'fool's gold.' I'm afraid you'll have to keep looking."

An afternoon full of false alarms left Angela and Paul pretty disappointed. "This search has led us nowhere," groaned Paul as he trudged down the trail. But Dad pointed out that while they never found the mine, they did hear a great story to share with their friends back home. As the Jordans arrived back at the trailhead, the sun was setting, and Paul turned to take one last look up the trail. There, in the fading afternoon light, he saw something sparkling among the towering rocks.

"We'll come back to Big Bend soon," he said firmly, "because I'm going to find that Lost Mine!"

Check In What part of the Lost Mine legend do you think might be true? Why?

GENRE Opinion Piece **Read to find out** about opinions on the fossils in Big Bend.

Where DINOSA Once Roame

Imagine a time when monsters roamed Earth. Picture gigantic crocodiles with enormous teeth, or huge dinosaurs with necks longer than five fifth-graders standing on each other's shoulders. Do you think Big Bend looks like a place such creatures might have lived? If so, you are right.

Big Bend's fossil record reaches back 100 million years, even before the time of dinosaurs. It includes fossils of plants, fish, amphibians, reptiles, dinosaurs, and early mammals. These fossils offer hints about prehistoric life and its **extinction**. When an animal or plant becomes extinct, it can no longer be found alive anywhere on Earth. By studying Big Bend's fossils, we can find clues to what life there was like long ago.

URS
d

by Dennis Fertig

> Based on this diagram, how would you describe the size of a SuperCroc?

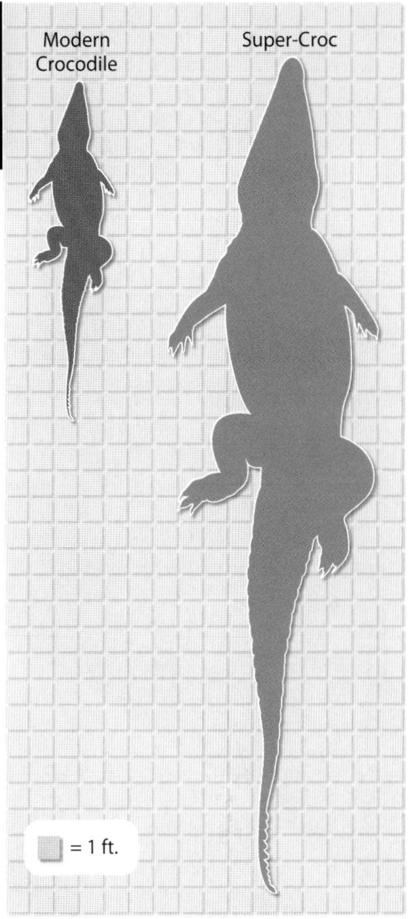

THE SUPERCROC

More than 90 species of dinosaurs have been unearthed at Big Bend, but one of the park's most famous finds is not a dinosaur at all. It is a terrifying, dino-eating crocodile nicknamed SuperCroc. Its official name is Deinosuchus (DINE-oh-SUE-kuhs), which means "terrible crocodile."

SuperCroc was up to 50 feet long. Its massive jaws were powerful enough to make a feast of any unwary dinosaurs that ventured too close to its swampy lair. How do we know SuperCroc was a dino-eater? Scientists have found lots of dinosaur bones in Big Bend covered with scars made by SuperCroc's six-inch teeth.

TURTLE VS. SUPERCROC?

Paleontologist Steve Wick's job is hunting for fossils in Big Bend, and it isn't easy. Temperatures there can reach 115 degrees, and off-road vehicles aren't allowed. Steve spends a lot of time hiking the hot desert in search of creatures from long ago.

During a scorching hike in 2004, Steve noticed an area littered with sharks' teeth and crocodile bones.

A closer look revealed dozens of thick shell fragments sticking out of a nearby hillside. Steve carefully **extracted**, or took out, the bits of shell and pieced them together like a prehistoric puzzle. They formed the top shell of an ancient turtle.

Steve Wick (left) and another paleontologist dig out the fossil turtle shell.

The next morning, Steve returned with another paleontologist. They gently extracted the turtle's bottom shell and carried the 140-pound fossil several miles through the desert to their truck.

Once the shell was safely back in the lab, Steve examined it and discovered a huge crack—a wound that had healed before the turtle died. He also found tooth marks. Had this turtle escaped a SuperCroc attack? We'll never know. Still, the turtle was quite a discovery: a rare, newly discovered species, and yet another treasure from Big Bend.

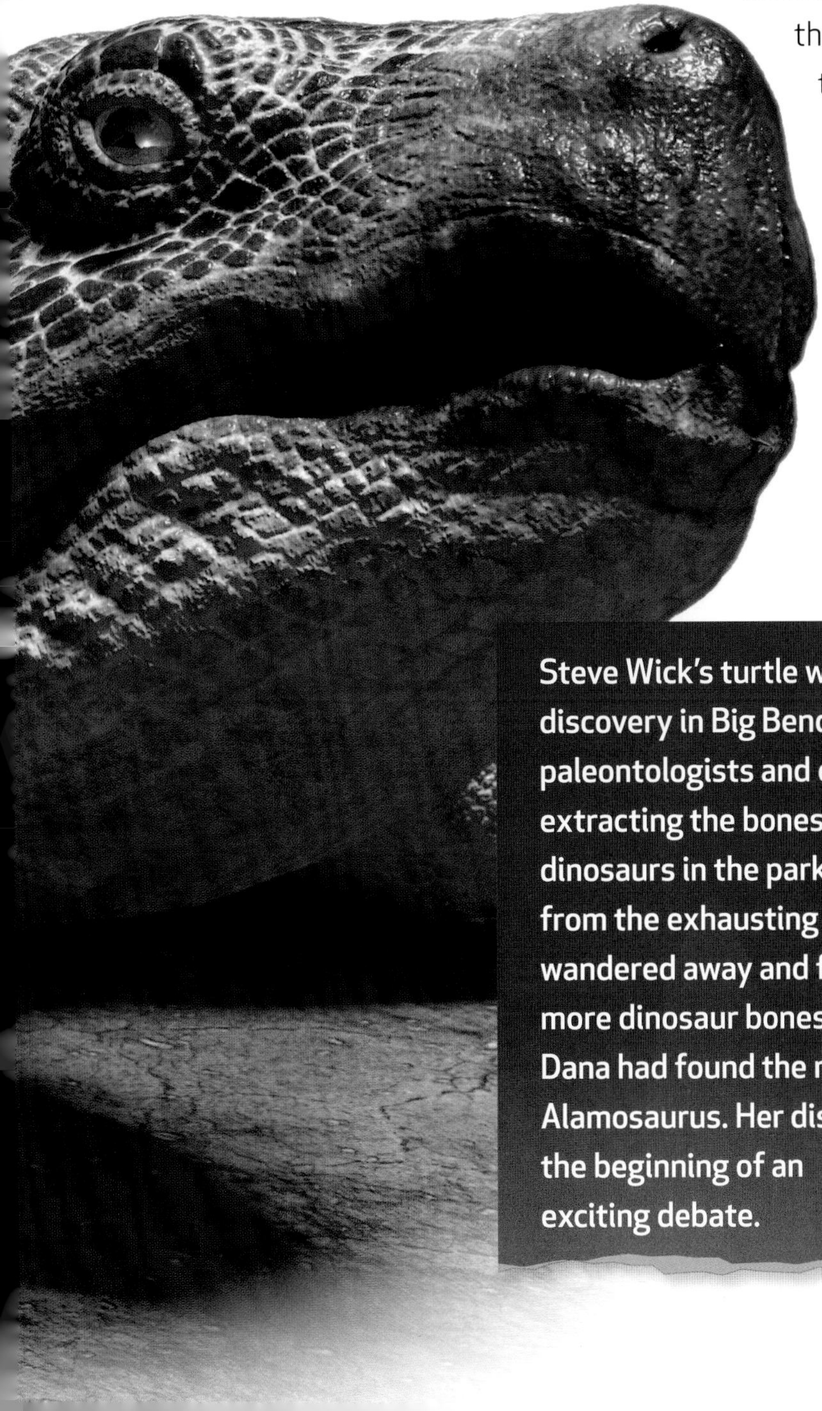

Steve Wick's turtle wasn't the only recent discovery in Big Bend. In 1999, a group of paleontologists and college students were busy extracting the bones of some young, small dinosaurs in the park. While taking a break from the exhausting dig, student Dana Biasatti wandered away and found herself gazing at more dinosaur bones. These were enormous! Dana had found the neck bones of an adult Alamosaurus. Her discovery was the beginning of an exciting debate.

Read on

The name Alamosaurus comes from Ojo Alamo, or Cottonwood Spring, in New Mexico. That is where the first Alamosaurus fossils were found in 1922.

BIG BONES IN BIG BEND

The Alamosaurus Dana Biasatti had found in the rock was a **sauropod** that lived about 70 million years ago. Sauropods were dinosaurs that ate only plants. They had long necks and tails and very small heads. They are the largest known dinosaurs.

The huge neck bones from Dana's dinosaur were really heavy—some weighed nearly 1,000 pounds! Neck bones that big probably came from a dinosaur that weighed 50 tons and measured 100 feet long.

These Alamosaurus fossils were an important discovery, but they presented a sticky problem. What should be done with them? Paleontologists were eager to study them and wanted to move them to a lab. Museums hoped to put them on display. But there were others who wanted the bones to stay right where they were. They said the fossils were part of Big Bend and should not be removed.

The park decided to move the fossils to a museum, which led to yet another problem. Few vehicles were allowed in the park. How could they get something the size of these enormous Alamosaurus fossils out of Big Bend? The answer? Helicopters.

In 2001, the Alamosaurus bones were airlifted out of Big Bend and brought to the Dallas Museum of Natural History, where they were cleaned and studied. Eventually, they inspired an exhibit at a new Dallas museum, the Perot Museum of Nature and Science.

> Paleontologists had to extract and prepare the fossils carefully for their journey.

A Big Bend Debate

The Alamosaurus bones sparked a debate about what to do with important fossils found in national parks. Should they be left where they are found, or should they be taken to museums and universities to be studied?

Leave them where they are. Many people believe nothing, not even special fossils, should be removed from a national park. They argue that national parks exist to protect precious natural resources, including fossils, and scientists could be allowed to study the fossils right where they are found.

Some local businesspeople suggested that the bones could become popular tourist attractions if they were left in the park. The bones could bring lots of visitors and their money to the region surrounding the park. They wanted to see an amazing exhibit built around the fossil site.

Remove them. Paleontologists agreed that fossils should rarely be removed from a national park, but they argued that Alamosaurus and similar fossils are very important finds. They were concerned that keeping these fossils in place might allow erosion to destroy them over time.

They didn't want to build an exhibit around the bones. They argued that it would bring visitors into previously protected park areas. Some people might even try to damage or destroy the bones.

> Scientists use the fossils they find to build models of dinosaur skeletons. These models help them learn more about dinosaurs such as Alamosaurus.

Reach a solution. Everyone involved knew the bones were important and wanted to protect them in the best way possible. Those who wanted to keep the bones in Big Bend felt that the fossils could be safely studied right where they were found. Those who wanted to remove the bones believed that the best plan was for scientists to study and preserve them in museums or at universities.

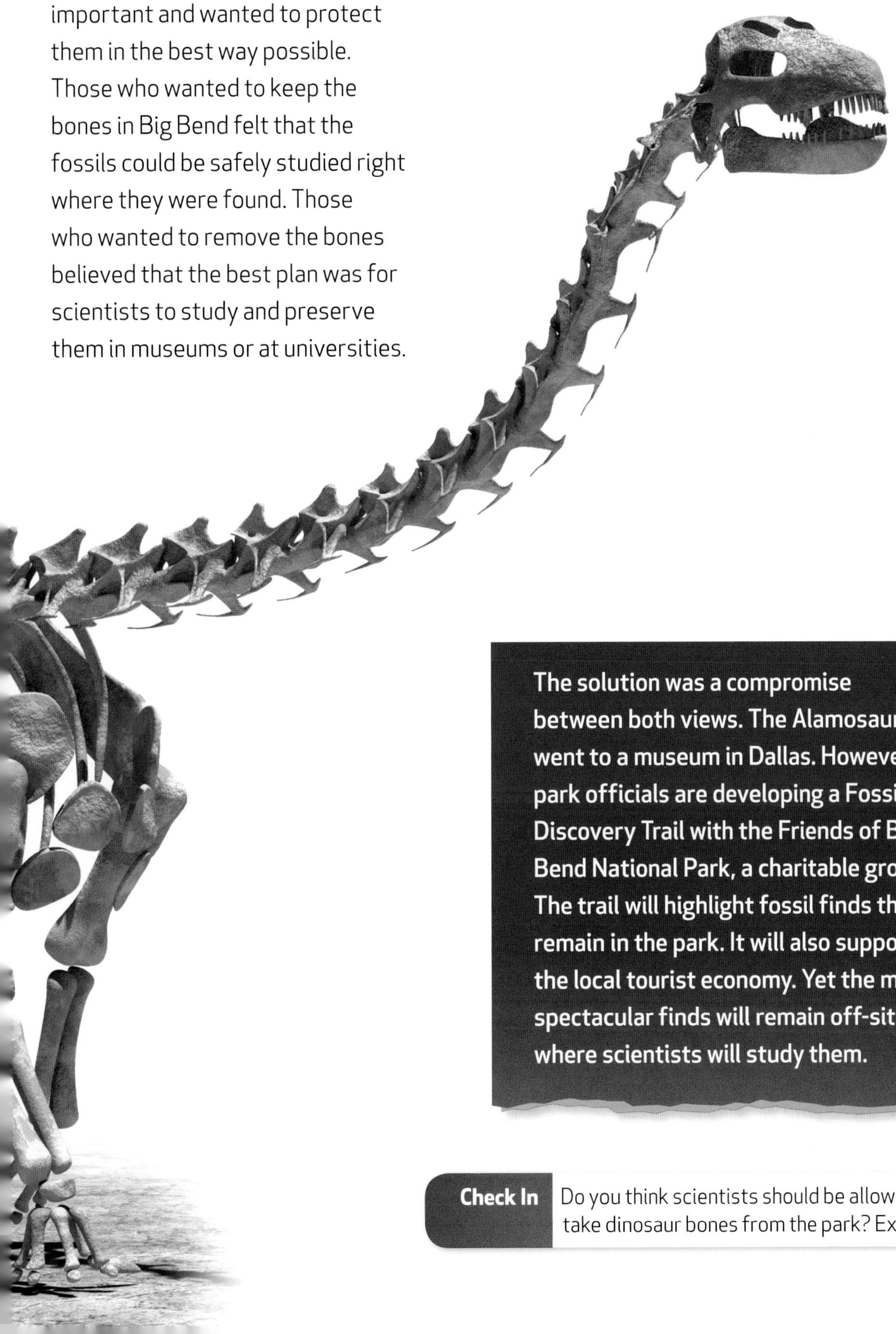

The solution was a compromise between both views. The Alamosaurus went to a museum in Dallas. However, park officials are developing a Fossil Discovery Trail with the Friends of Big Bend National Park, a charitable group. The trail will highlight fossil finds that remain in the park. It will also support the local tourist economy. Yet the most spectacular finds will remain off-site where scientists will study them.

Check In Do you think scientists should be allowed to take dinosaur bones from the park? Explain.

Discuss

1. What connections can you make among the four selections in this book? How do you think the selections are related?
2. Why was the Big Bend ecosystem named as a biosphere reserve? How does that help protect and preserve the different habitats within the park?
3. What can bird-watching teach you about the different habitats in Big Bend National Park?
4. How would you summarize the debate about taking dinosaur bones out of Big Bend? Was the solution a good one? Explain.
5. What do you still wonder about Big Bend National Park? How can you learn more?